W9-BYL-492

BETTY & FRIENDS

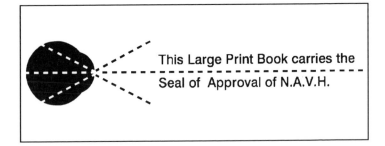

This Large Print Book carries the
Seal of Approval of N.A.V.H.

BETTY & FRIENDS

MY LIFE AT THE ZOO

BETTY WHITE

THORNDIKE PRESS

A part of Gale, Cengage Learning

GALE
CENGAGE Learning™

Detroit • New York • San Francisco • New Haven, Conn • Waterville, Maine • London

GALE
CENGAGE Learning®

Copyright © 2011 by Betty White.

Thorndike Press, a part of Gale, Cengage Learning.

ALL RIGHTS RESERVED

While the author has made every effort to provide accurate tele-
phone numbers and Internet addresses at the time of publication,
neither the publisher nor the author assumes any responsibility
for errors, or for changes that occur after publication. Further, the
publisher does not have any control over and does not assume any
responsibility for author or third-party websites or their content.

Thorndike Press® Large Print Nonfiction.

The text of this Large Print edition is unabridged.

Other aspects of the book may vary from the original edition.

Set in 16 pt. Plantin.

LIBRARY OF CONGRESS CATALOGING-IN-PUBLICATION DATA

White, Betty, 1922–
    Betty & friends : my life at the zoo / by Betty White. —
Large print ed.
      p. cm.
    ISBN-13: 978-1-4104-4525-4 (hardcover)
    ISBN-10: 1-4104-4525-9 (hardcover)
    1. Zoos — United States — Anecdotes. 2. Zoo animals —
United States — Anecdotes. 3. Zoo animals — United
States — Pictorial works. 4. White, Betty, 1922– I. Title.
II. Title: Betty and friends.
    QL76.5.U6W45 2011
    636.088'9—dc23
                                                                    2011040112

Published in 2011 by arrangement with G. P. Putnam's Sons, a mem-
ber of Penguin Group (USA) Inc.

To Mary Uneda,
Tad's wonderful mother

This is my girl.

CONTENTS

INTRODUCTION

For openers, let me say that I realize there are those whose minds are irrevocably set against the entire concept of zoos and consequently see only negatives. They have every right to their opinion, and I respect it. As for me, however, I am a confirmed zoophile, and I particularly appreciate the positive changes that have taken place in the whole zoo community over the past few decades, and the critical role they play today in perpetuating endangered species.

My interest started early on, tagging

after my mom and dad, who went to the zoo often, not just to please their little girl but because they enjoyed the experience — some zoos, of course, more than others — and would have gone even if they never had me.

Growing up in busy downtown Chicago, neither of them had come from families even slightly animal-oriented, but both my mother and father were genuine animals nuts, and I am eternally grateful that they have passed much of that passion on to me.

It was from them I learned that a visit to the zoo was like traveling to a whole new country inhabited by a variety of wondrous creatures I could never see anywhere else in quite the same way. They taught me not to rush from one exhibit to the next but to spend time watching one group until I began to really <u>see</u> the animals and observe their interactions.

My folks also taught me to discriminate between the "good zoos" and those other places that displayed animals for all the wrong reasons and sent you home feeling sad. So it was inevitable that I became a lifelong card-carrying zoo groupie.

Wherever I travel, I try to steal time to check out whatever zoo is within reach. While I was in New Mexico doing a movie I enjoyed the Albuquerque Zoo. When in Atlanta I always spend time at both Zoo Atlanta (with its record twenty-two lowland gorillas) and the beautiful Georgia Aquarium. More on both later.

Chicago has two fine zoos — Brookfield and Lincoln Park, the latter of which is actually admission-free!

Back home, of course, there is my beloved Los Angeles Zoo and Botanical Gardens. No matter where it may be, I will never come away from a zoo visit without having seen something or learned something to remember.

How many zoos are there in the world? Take a guess. If you say hundreds, you fall short. Worldwide, there are more than a thousand zoos. Each year, over 175 million people visit the zoos and aquariums of North America, which is more than all those attending professional sporting events combined.

Zoos trace back to somewhere around 2500 BC, when the nobility began building private collections of exotic animals for their own amazement. Of course, human nature being what it is, the idea soon evolved into wanting a better collection than somebody else.

"My zoo is bigger than your zoo!"

"I want one of everything!"

"If an animal dies, I'll just go out and catch another one."

Unfortunately, that general attitude persisted more or less through the centuries, until, at long last, the law of diminishing returns began to dawn. High time.

Several zoos in this country claim to be the "first" American zoo — Central Park Zoo in New York, the Lincoln Park Zoo in Chicago, and so on. But the first zoological garden (which is the formal term for a place where live animals are kept, studied, bred, and exhibited) was the Philadelphia Zoo. It opened on July 1, 1874. The zoo would have been built sooner, since the actual charter establishing the Zoological Society of Philadelphia was signed fifteen years earlier, in 1859. It was the Civil War that put building plans for the zoo on hold.

As of 2012, there are two hundred and twelve accredited Association of Zoo and Aquarium institutions in North America. Because of the burgeoning human population, wild habitats around the globe continue to get smaller, or even to disappear altogether. Increasingly, the good zoos find themselves taking on the role of "protectors" — or better yet, "conservers" — rather than merely "collectors" of wildlife.

Through the years I have visited many zoos, and I would love to introduce you to some of the friends I've made along the way. As you have gathered, zoos are incredibly important to me. In this book, I hope you'll find out why.

Come along.

■ ■ ■ ■

Zoo
Good
Deeds

■ ■ ■ ■

California condor

I'm often asked how I got so involved with the Los Angeles Zoo. In 1974 I became a trustee of the Greater Los Angeles Zoo Association (GLAZA), the nonprofit support organization of the L.A. Zoo.

But it started long before that.

Prior to the current zoo being built (it officially opened in 1966), the L.A. Zoo was a small hillside menagerie in Griffith Park. I used to go up there by myself to wander around and visit with the animals. I'm not well, as you know by now.

It was hard to believe that a city like Los Angeles would have such a poor zoo. I've never been one to stand outside and join critical demonstrations — I wanted to get inside and get involved. It turned out to be one of the happiest choices of my life.

Over time, the zoo has come a long way. The first major project was our chimpanzee exhibit. We had a large chimp troupe — fourteen animals in all — displayed together in what had been an old bear grotto. At night they slept in small cages in a limited area. The need for help was painfully obvious.

Blessedly, the public voted a big yes on a proposition that gave us the much-needed wherewithal to address the problem. We even consulted Dr. Jane Goodall for some advice on the

design of the exhibit. No one under-
stands chimps and their needs better.
That is when Jane and I met and our
lasting warm friendship first began.
She is so special.

At last we got a state-of-the-art
Chimpanzees of Mahale Mountains
exhibit built for our animals, who
now have a large green two-story environment with good night quarters,
and they enthusiastically make the
most of every inch of it.

Next came the Red Ape Rainforest
for our orangutans, followed by a great
new home for our gorillas. We worked
hard to raise the daunting funds each
time for these projects. At one time
I probably could have sold my body,
but that market disappeared.

This all goes way beyond just building more interactive, spacious, and

attractive exhibits that benefit the animals and visitors alike. Good zoos are critical to conservation. The information we have gained regarding disease, breeding, and health issues from our captive animals we can apply toward helping those in their natural habitats. Zoos are doing extraordinary work with the wild populations of endangered species around the planet, as well as raising awareness of the situation in their native countries.

If it weren't for zoos, there are many animal species that would be extinct today. The California condor is a great example. In 1982 the number of California condors in the wild was less than two dozen. That's when the L.A. and San Diego zoos teamed up to establish a condor breeding program, bringing wild condors into the safety of the zoo habitat to breed and replenish their numbers, then releasing them back into the wild. Today you can see California condors flying in the skies above California, Arizona, and even Baja, Mexico.

Or take the critically endangered Sumatran rhinoceros. It's believed there are less than two hundred living in the wild, and fewer than ten in captivity. In 2001 the Cincinnati Zoo was host to the first live birth of a Sumatran rhino in captivity in one hundred and twelve years. His parents had been rescued from southwest Sumatra, where deforestation has decimated the rhinoceros population. When his mother conceived again, Andalas was transferred to the L.A. Zoo, where he became a star attraction. But in 2007 the L.A. Zoo sent him to a sanctuary in his native country to breed with other Sumatran rhinos in the hopes of building the devastated rhino population.

Why do I keep thinking of Noah and the ark?

BLACK BEAR

One of the other great deeds zoos do is save animals that have been abused or orphaned due to violence or environmental disasters.

This guy was a forest-fire refugee, rescued as a cub and brought to the zoo.

Andalas, the Sumatran rhinoceros

There are five different species of rhinoceros — the white rhino, black rhino, Indian, Javan, and Sumatran — all of them critically endangered.

On September 13, 2001, Andalas, the first Sumatran rhino born in captivity in more than one hundred and twelve years, became a worldwide news sensation. His mother, Emi, had been a dear friend of mine when she lived at the L.A. Zoo. I would often go visit with her before the zoo opened in the morning and she'd sing me little squeaky songs. I missed her terribly when she was moved to the Cincinnati Zoo for breeding purposes.

Breeding in captivity can never be a substitute for protection in the wild, but at this critical point in fifty million years of rhinoceros history and with the global Sumatran population down to no more than two hundred wild animals and only ten in captivity, we *must* keep this species from extinction.

One of the rarest of the five rhinos, the Sumatran is the only one with patches of short, stiff, red-brown hair, which helps keep mud caked to their bodies to cool them and ward off insect bites.

At birth, Andalas weighed in at seventy pounds, and by his first birthday he was up to nine hundred forty. When he was almost two, he was weaned and transferred to the Los Angeles Zoo to make room for a new addition since his mom, Emi, bless her, had conceived again! In California our boy grew to more than one thousand six hundred pounds, and by five he had reached his full adult size.

It was recommended by the Global Management Propagation Board for Indonesian Rhinos that the fully grown Andalas be moved to join the breeding program at the Sumatran Rhino Sanctuary in Sumatra.

It was a long and carefully monitored journey. Andalas was put aboard a special plane in the cargo area but he did not travel alone. By his side in the cargo area were his keeper, Steve Romo; our Los Angeles Zoo veterinarian, Dr. Curtis Eng; and Dr. Robin Radcliffe from the Veterinary

Rhino Foundation.

The round-the-world flight went well, and after Andalas cleared customs in Jakarta, a twelve-hour road trip brought him to the sanctuary, where a special "boma" — a pasture and paddock area — was waiting for him. A sixty-three-hour journey was completed.

Dr. Eng and Dr. Radcliffe stayed with him for a week to be sure he adjusted and then a two-month quarantine followed, during which this invaluable creature was monitored hourly by veterinarians and keepers.

After these months he was released into a big paddock and gradually introduced to two young females, Rosa and Ratu.

Andalas has adjusted well to life in Sumatra. Adapting to the heat and

Emi

humidity of the Indonesian rain forest, he shed some of the orange hair he had grown in the United States.

Andalas is now ten years old, and everyone has high hopes that he will successfully breed with one of the three females at the sanctuary.

If that happens, our Emi — my Emi — can take much credit for helping to save her whole species.

THE KEEPERS

The relationship between the keepers and their animals is nothing less than phenomenal — the understanding and the communication. The trust and the friendship. And there's just no limit to what the keepers will do for their animals in the face of a threat. In 2007 there was a fire in the L.A. area, and the keepers just came like magnets from all over to protect their charges — not because they were required to but because they said, "These are my kids!"

And they're so generous, wanting

to tell you about their kids. They're extraordinary people.

I've always thought you can tell by someone's hands if they truly "love animals" — the way they pet or don't pet them. You see it with the keepers in spades. To be honest, I used to judge boyfriends that way — how they greeted my dogs decided exactly whether there was going to be a next date!

Early one morning I was in the zoo and one of the keepers was out there shoveling the exhibit. A father and son stood nearby. The father said, "You see? If you don't do your homework you'll end up doing that!"

What the father didn't know was that this keeper was a Ph.D.! We have several keepers with advanced degrees among our wonderful staff.

■ ■ ■ ■

THE EYES
HAVE IT

■ ■ ■ ■

Jaguar

The most fascinating thing to me about animals starts with their eyes.

When I was a kid, I'd go to the San Diego Zoo and I'd watch the baby gorillas and wonder what on earth they were thinking. I'd sit there for hours and look at them, and they'd look at me, and I think that's when my interest in animals was first ignited.

I still do that — stare at an animal, wondering what's going on behind those eyes. It doesn't matter the species. You realize these are the same eyes that protect them and keep them alive, the eyes on which they depend so much.

Crested capuchin

Chimpanzee

Meerkats

Squirrel monkeys

Serval

Golden-cheeked gibbon

■ ■ ■ ■

STANDOUTS

■ ■ ■ ■

My preoccupation with animals is an open secret, and I am often asked which of them all, domestic pets aside, is my favorite. I'm always stuck for an answer.

Where does one begin to choose? Gorillas? Elephants? Big cats? Bears? Rhinos? The list goes on. Yes to all of the above, plus any others you might name.

Perhaps a simpler approach would be to ask which animals I don't find fascinating, and I could honestly say, "None." The only ones I find disappointing on occasion are the two-legged type, but they are still interesting, nonetheless.

Gita

GITA

There are individual animals, of course, who stand out. Who take your heart and won't let go. For me, there is one who will always rise above the rest, not only in stature but in every way. That was our beloved Asian elephant Gita. Years ago, when I was first introduced to this big girl, I said, "Trunk up, Gita." She obliged, and I stood on tiptoe to slap her tongue. Somewhere along the line, I had learned that elephants like having their tongues slapped, and it really worked with Gita. She reacted as though I had found the right language to communicate with her, and our long friendship began. And that became our traditional greeting.

It was in subsequent visits to the zoo, when they did have contact elephants on-site (those you can touch), that I found that Gita wasn't the only elephant who liked to have her tongue slapped. I couldn't help wondering where they learned that.

But Gita was truly exceptional. I was amazed at her gentle disposition and human awareness/curiosity. I was delighted that I could get my hands on her — which is a privilege, these days, particularly.

Early each Saturday morning, the keepers would take Gita on a long, rambling walk around the grounds before the zoo opened, and it was my privilege to be invited to go along, several times. Those are the memories etched in my mind. I remember the first one and how excited I was when I arrived at 8:00 a.m. to find Gita lying on her side to be swept off. Then she got to her feet to be hosed down. Then the keeper and Gita and I took a long stroll all through the zoo. Now and then, there would be a quiet "Steady" from the keeper accompanying us, not because Gita had done anything wrong, but rather as a word of approval and encouragement. She was just taking a walk with her friends.

The walks provided great exercise for her and great keeper-elephant confidence-building as well. Back in 1966, when the zoo had moved down the hill to its new location, Gita had not only walked there, but she had led some of the other animals in a make-shift parade! It was probably a couple of miles.

Gita was the matriarch from then on.

So on those morning walks I was so thrilled to be dealt in. We'd all just walk out and go around the zoo. We'd go by the chimpanzees and they'd get frantic, moving around, getting sooo busy — There's an elephant coming! An elephant! But Gita paid them no mind.

She would stop and try to eat things — wrap her trunk around a young sapling and be just about to pull when the keeper would say — "Ahhh, no, Gita!" And you could see it on her face, Aw, shucks. Okay. And we'd move along.

Gita and Billy

Billy at play

Whenever I got to walk with Gita, I was in heaven. Walking beside her, I was always amazed that I didn't hear a footfall. Gita would mumble or make little squeaks as if to join in the conversation, but those big feet were silent as they hit the ground.

Gita was forty-eight years old when she left us, but she will stay in our hearts forever. Yes, Gita, dear, you are still my very favorite of all.

ELEPHANTS

The new pachyderm exhibit at the L.A. Zoo is another long-term dream come true. We not only went from one acre to four acres but included beautiful hilly terrain that our elephants are enjoying.

Elephants are not only the largest land animals, but they may also be the most recognizable on the planet with those unique trunks. We tend to take them for granted, but these days they are in real need of help.

The Asian elephant is critically endangered due to being illegally hunted for its ivory tusks, plus the destruction of its habitat. Those who demonstrate against any elephants being kept in zoos seem unaware of these statistics and the opportunity to educate the public about these animals. In zoos like the one in Los Angeles, the majority of our visitors will never be able to travel to locations to see elephants in the wild.

There are two elephant species — Asian and African. The differences begin with the ears. The African elephant's ears are enormous and, oddly enough, shaped like the continent of Africa. The Asian elephant's ears, while much smaller, are shaped like India.

And they use their trunks differently. The African elephant uses his like fingers and picks things up as you and I would with our thumb and forefinger. The Asian elephant picks things up by curling that trunk around them, but is so adept he can pick up the tiniest item.

African elephants have four toes on each foot, Asian elephants five. Watching an elephant walk, with legs like moving tree trunks, it is difficult to realize that these gigantic animals are actually walking on their toes. Considering the bone structure inside that huge foot, the way an elephant walks is comparable to the way a woman walks in high heels.

ORANGUTANS

High on my list of most intriguing animals has always been the orangutan — "man of the forest" — who is so deeply intelligent yet so different from his fellow primates: the chimpanzee or gorilla, or human. Whatever zoo I visit, I often come away with an indelible impression of some orang I just met. As a case in point, to this day, even after several years, I have a fond and vivid memory of a lady orang with whom I played a serious trading game behind the scenes at the Columbus Zoo in Ohio.

It began when she handed me a small stick under the gate of her enclosure. I looked at it carefully and gave her a small stone in exchange. She examined it closely, then found a piece of straw and, looking straight into my eyes, handed it to me. Your turn!

This wordless back-and-forth went on for perhaps ten minutes, and it was a personal, private one-on-one experience I shall never forget. I have since discovered that trading is a favorite game with many orangutans, but she was special.

Most primates, humans included, are group-oriented — social animals for whom the family unit is a way of life. By contrast, the orang spends most of his time alone, living in the trees of his native Borneo or Sumatra. The rain forest continues to be burned to make room for palm-oil and pulpwood plantations, and the tragic result is that thirty to fifty percent of those trees have been lost. This leaves wide, treeless swaths that isolate the arboreal orangutan and cast a shadow on his future in the wild.

Orangutan mothers nurture their babies for eight to ten years — longer than any other mammal except humans! They teach them everything — where to find food in certain trees at certain times of year; where to find shelter. Orangs are incredibly smart and have terrific memories. This compounds the tragedy of deforestation — those orang children remember the location of trees their moms showed them, which have now disappeared. There are photos of

Minyak

these ten-year-old orangs just staring at these treeless swaths, not knowing where to find food.

Properly maintaining an arboreal animal who prefers a wide-ranging aloneness can present quite a challenge in captivity. Our Red Ape Rainforest is grass-covered, with a mesh canopy affording the orangs the opportunity to climb to the top of the habitat and survey their surroundings (zoo guests and their elephant neighbors). It also has tunnels for exploring and areas for seclusion, in case one wants to be alone.

Though they seem to enjoy socializing in the day, at night they sleep in separate spaces, giving them much-needed alone time, which they seem to appreciate.

The Red Ape Rainforest exhibit has

a complex design, but these guys are so worth the effort.

Bruno being silly

BRUNO

Bruno, our three-hundred-and-eighty-pound redhead, is the largest in our orangutan group. Minyak, our other male, is smaller, weighing in at two hundred pounds, although both orangs are around the same age — thirty-two years old.

Bruno and I have what I like to think of as a unique rapport. When I stop by to talk to his keepers, all of whom are female (growing up, it was more than rare to see a female zookeeper), there is always a little special Bruno time. I'm sure the girls are just being kind by saying Bruno is different with me, but I lap it up.

Every time I go to visit, I ask if this is the day I get to take Bruno home with me. I just love that sweet guy.

One morning I was there very early and was allowed to stand by a section of behind-the-scenes fence. Bruno was way back at the other end of his habitat, sprawled out on the grass, basking in the morning sun. I called him — just a friendly hello, not expecting any response. You could almost see the wheels go around in his head: Is it worth the effort?

After a moment, he pulled all three hundred and eighty pounds together, got up, and slowly made his way down the whole length of the exhibit to where I was standing. He plopped down and pushed his lips through the fence so I could rub his nose. It wasn't food-oriented, I had no reward for him — it was strictly social contact.

I was the one who got the reward.

MINYAK

Though Bruno and Minyak are the same age, we don't put the male orangutans together. Minyak came to us as a last resort — he had an infection in his air sac. Air sacs are normal in orangutans, but when he came to us he was desperately sick, and the vets didn't know if he'd make it. But with treatment and monitoring, he's made a great recovery.

Minyak's genetic disposition is incredibly valuable because both his parents were born in the wild. This means he's unrelated to any of the orangutans already at the zoo, which is terrific for diversity. Minyak is the father of Bosco-Berani, and we are anxiously awaiting the birth of Minyak's second offspring, which, fingers crossed, should be born by the time this book comes out!

Loner though an orang may be, he makes a notable exception at mating time, whether in the wild or in the zoo.

Bosco-Berani, aka "Bera"

Here Bruno conducts a very effi-
cient courtship.

Just ask Kalim. . . .

KALIM

Female orangutans give birth only once every eight to ten years. That makes sense, since mother orangutans take care of their babies longer than any other land mammals except humans.

Our Kalim was so thrilled with her firstborn that she would come to the glass viewing area and hold little Bera up for visitors to see.

Isn't this the most beautiful baby EVER?

I recently visited Kalim, who's pregnant again (Minyak is again the father), and she and the zookeeper had a wonderful trading game going on. The keeper had grapes, which she was mostly feeding Bruno, but Kalim got brave and came to the door as well. However, Kalim decided to "pay" for her snacks by passing twigs to the keeper, who would pass back a grape. Fascinating to watch!

They say if you put an empty box in a gorilla cage, he'll tear it up. If you put it in a chimpanzee cage, he'll stomp on it and tear it to pieces. If you put it in an orangutan cage, he'll fold it into some useful structure and find a way to make it part of his nest. They're smart as whips!

Kalim holding Bera

Lionel

LIONEL
AND COOKIE

Lionel and Cookie were brought to the zoo from the Wildlife Waystation Sanctuary in Angeles National Forest in 1997. They were inseparable.

We lost Lionel recently. He was twenty-three years of age. In the wild, male lions would be lucky to live to twelve years old, so he had a good run, much as we miss him.

If Cookie were younger, we'd try to find her a mate. But she's twenty-two now, and doing just fine for a single gal.

Of course, the female lion, who does most of the work, is a thing of beauty herself.

Cookie

They were a very happy, if relaxed, couple.

HOWLER MONKEYS

This photo shows the beautiful color phases of these fascinating South American animals. In this case, the black howler monkey is male, and the female, tan. What I <u>wish</u> we could demonstrate here is the incredible sound they are making.

Early in the morning, shortly after dawn and well before the zoo opens, you can hear the howler monkey song. It is such a piercing sound it carries all over the zoo.

A nice way to start the day.

Smokey and Dottie

Our Jaguars

Meet Smokey and Dottie, two color phases of the same species of jaguar. Look closely at that beautiful black face and you will see that he, too, has a pattern of spots showing dimly through the black. Catch him in bright sunlight and spots are discernible over his whole body.

Smokey and Dottie are gorgeous, but it's the training of our other jaguar, Kaloa, that fascinates me. The keeper has established a marvelous rapport with him. When she sits down on the ground outside of but close to the enclosure, the jaguar knows it's training time, and he comes to sit as close to her as he can get. She has him go through various procedures to get his reward, a treat that she is able to give him — by hand — through the wire. No one else can do this with him. For instance, on her request he will slip his paw out under the gate or press his shoulders against the wire that is separating them.

This is wonderful to witness, but it is not done as a performance. Should Kaloa need medication at any time, he will put that paw out when the keeper asks, and she can give him an injection without his having to be tranquilized.

Many zoo animals these days are trained to voluntarily offer whatever body part is in need of treatment

Smokey

without having to be put under. Even the small hyperactive monkeys get the idea.

Much less stress is involved. Just cooperation with a trusted human friend. Classic zookeeper accomplishment.

Reggie, the truant alligator

REGGIE

In Los Angeles, for months there was a startling sight: an alligator, certainly no California native, was swimming in Harbor City's Lake Machado, and authorities couldn't catch it. It's likely he was an exotic pet someone unwisely adopted and then couldn't care for, so he was just released illegally into the lake. Exotic pets are an epidemic. In terms of black-market moneymaking, they're probably second only to the drug trade.

Reggie became a big local celebrity and was finally trapped in a coordinated effort between the City of Los Angeles Parks and Recreation staff and the L.A. Zoo's curator of reptiles and amphibians and brought to the zoo.

For every Reggie a zoo is able to take in, zoos across the nation receive dozens of calls from people who want to find homes for their snakes, iguanas, and in some cases bears and monkeys. In most cases, the zoos are already at capacity.

How many times do we have to repeat: <u>Exotic animals must not be kept as pets?</u>

Sifaka

SIFAKA

Sometimes you almost wish you could sit down and have a conversation with these animals.

LINA

We had a wonderful gorilla named Lina at the Los Angeles Zoo. I used to hold her as a baby, and we just adored each other. When she got older, she learned to blow kisses. I'd stand outside, and Lina would blow kiss after kiss. I couldn't get enough of it.

Years later, I went to New Mexico for a movie, and of course I <u>had</u> to steal time to get to the zoo in Albuquerque, which is lovely. It has this all-natural environment — called a "BioPark zoo" — with steppes and terrain the animals experience in the wild.

They had this wonderful gorilla exhibit. When we arrived, I noticed that one of the gorillas seemed to be blowing us kisses.

"Oh," I told the zookeeper, "our Lina used to do that!"

She turned to me and said, "That <u>is</u> Lina!"

Years ago I made friends with a very special little girl, Lina.

Here's Lina, in Los Angeles.

. . . And later in Albuquerque.

BEETHOVEN

I have a very dear friend who lives in Atlanta. He just happens to be a beluga whale.

You can't meet someone like Beethoven and never see him again.

So I flew myself out to revisit the Georgia Aquarium recently. I know he didn't, but it was almost like he recognized me. He sped across his enormous pool to give me a welcoming kiss.

Dr. Warren Thomas with Caesar

CAESAR

Dr. Warren Thomas was director of the L.A. Zoo from 1974 to 1990.

A tremendous animal man, he actually helped deliver the very first baby gorilla ever born in captivity at the Columbus Zoo in Ohio. When he came to Los Angeles, we were ecstatic and hoped against hope that he would repeat the event. And the miracle actually happened.

When the time came for the birth, we not only had Dr. Thomas, but people doctors from the hospital downtown came to help deliver the baby male gorilla by cesarean section.

Of course we named him Caesar, and he became such a star.

And this is little Caesar

Caesar all grown up

OKAPI

Do you know what an okapi is? Many people will answer that question with a blank stare — they have no idea. Well, prior to 1901, no one else did, either, for that is when this gorgeous antelope was first discovered in the forested areas of Africa — primarily the Congo and Zaire — and it is the last large mammal in the outside world to be identified to date.

The okapi is a distant cousin to the giraffe, which shows in its long neck and small head. Its fur looks so close-cropped, but you could lose your hand in it. It's remarkable that such a large, and one might say <u>gaudy</u>, animal could stay hidden for so long; however, they are virtually impossible to see in the wild, given their speed, camouflage, and scarcity.

They are on exhibit and reproducing in several major zoos today. It's amazing to realize they have been known for only a little over one hundred years.

The L.A. Zoo received its first okapi in 2005.

I felt most privileged to get this up close and personal with Jamal!

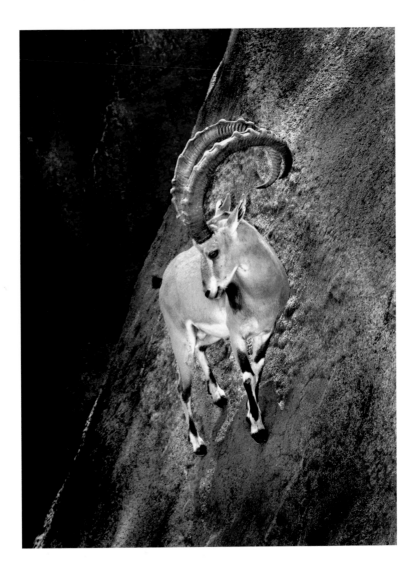

Nubian Ibex

This amazing photo captures the ibex's uncanny ability to scale walls — it looks like its hooves are on nothing! And that's practically the case. This is what they do to escape predators.

When you look at this incline in real life at the zoo, it is just that perpendicular — you can't believe anything could stand on it!

This photo won Tad Motoyama first prize in the Association of Zoos and Aquariums photography contest.

■ ■ ■ ■

FAMILY
MATTERS

■ ■ ■ ■

Chimpanzee family

The exciting zoo season is spring, when our babies begin to arrive.

In 1987 it was hard to imagine that this wide-eyed young-ster would grow into our imposing silverback, Kelly.

KELLY

Kelly is a twenty-five-year-old silver-back gorilla and the dominant male in one of our two separate groups of gorillas at the L.A. Zoo. Kelly weighs a daunting four hundred and fifty pounds, which is a far cry from when he was a newborn baby at the zoo in 1987.

Kelly and Evelyn were supposed to breed, but that didn't happen. Then it was hoped that Kelly would mate with another of our females, Rapunzel. A baby gorilla is a true crowd-pleaser.

We waited and waited, but nothing was happening, and we wondered if there would <u>ever</u> be chemistry between Kelly and the ladies.

In 2003 we began building our new gorilla exhibit. So that the animals wouldn't be upset during the construction, we sent them on loan to the Denver Zoo.

Well, I don't know whether it was the altitude or wine and soft music or what, but Kelly and Rapunzel soon hit it off and she had not one but two mile-high babies during their Denver stay. Her second baby, Glenda, was about two years old when she came back to the L.A. Zoo with her mother and father. She turned six in May 2011.

Evelyn, meanwhile, never bred, but she has become our beloved elder stateswoman. At thirty-six years old, she gets grudging respect from the younger girls, and even from Kelly.

Here is Kelly in all his glory at age twenty-one.

Evelyn and Glenda

So many of the males can't be trusted with a baby . . . unlike Kelly. Here he is enjoying gentle play with his daughter, Glenda.

Glenda is not above using her mom, Rapunzel, as an accomplice.

Evelyn

Kelly: I can do it.
Glenda: Well, I can, too!

It was a great surprise to discover that apes actually use tools. Here's Rapunzel proving it.

CAMEL FAMILY

Years ago at the L.A. Zoo, we had a group walk as a fund-raiser. When we got to the camel exhibit, one of our two females — hugely pregnant — was starting labor. No way could I leave until that baby was born, so I pulled out of the walk and just hung out with the keepers. It was almost an hour before the mother finally dropped that little package to the ground.

As with most hoofed animals, a baby must get up and be able to run as quickly as possible so as not to risk being trampled or taken by a predator.

This little guy kept struggling to get those long legs in line to support himself. He would no sooner get one set when it would fold before he could get the other three in order.

Finally — at long last — he made it to his feet and stood by his mother. He was teetering a bit, but he was on all fours.

Just then our other female camel wandered over like an affectionate aunt and, as if to say, Oh, what a beautiful baby! she touched him lightly with her nose.

SPLAT!!!

Down he went, and had to start the whole process all over again.

These handsome fellows are Bactrian camels, characterized by the two humps. (It's the dromedary that has only one.) The animals pictured here are exceptionally beautiful examples, with their firm upright humps. Often those humps will sag and tip over in adults. I know the problem.

Bactrian camel

Wild Bactrians come from the inhospitable steppes of Central Asia, although most of these creatures are currently domesticated. Just think how long and how closely they have been working for humans. In some areas, people couldn't make it without them.

MARKHORS

Markhors are a type of wild goat found in the mountainous regions of Pakistan and Afghanistan and Uzbekistan. There are fewer than seven hundred and fifty of these animals living in the wild today.

Mommies
and Babies

GIRAFFES

Giraffes, indigenous to Africa, are the tallest mammals on earth. The males can get up to eighteen feet tall, the females up to fifteen feet! Challenges begin early for baby giraffes, who face a near-six-foot drop to the ground in their first seconds of life.

Getting those long legs untangled and struggling to her feet, our little Harriet found a safe shelter under her protective mother.

Those first few moments are always anxious ones, and it is a great relief when the newborn stands and finally stretches that little neck to nurse. Remarkably, within hours, these nearly six-foot-tall newborns are running around under their own steam!

TIGERS

A little horseplay — or tigerplay.
  One thing can lead to another . . .

. . . or even three others.

Our Sumatran tiger Lulu is pictured with her second litter, and she has been a terrific mother.

This isn't always the case. Some years back we had a tiger who wanted no part of motherhood. One morning she gave birth to three babies and would have nothing to do with them — she left them completely alone.

The keepers knew the tiny kittens couldn't make it on their own and time was of the essence. One keeper managed to divert the mother's attention long enough for the other keeper to scoop up the kittens and rush them to the zoo nursery.

As luck would have it, I happened to be there at the time, talking with a nursery keeper about a zoo special I was preparing to produce. By now the kittens were in real need of help, as they were cold and fading fast. We each took one of these little creatures, which just fit into our hands, and began to rub — hard. We rubbed and rubbed until, at last, their color began to normalize and a hasty formula could be put into baby bottles to get them some help for inside as well. Thankfully, all three made it.

In retrospect, I've always been amazed by the fact that although tiny puppies don't look anything like the dogs they will grow into, and even lion cubs are certainly different from the adult animal, these little guys were absolute replicas of Mom and Dad. They were already perfect tigers — only in miniature.

Sultan and Rani

MOUNTAIN TAPIR

The highly endangered mountain tapir is native to South America. A baby tapir looks very much like a rattlesnake watermelon with stripes for camouflage that soon disappear.

Yellow-footed rock wallaby and joey

Colobus monkey and baby

This is the red river hog from Central Africa, with three of her litter of five.

Flamingo with her chick. As he grows up, his feathers turn pink.

White-faced whistling duck

Here's Cleo and her son, Kelly.

Chacoan peccary and babies, found in Paraguay, Bolivia, and Argentina

Our beautiful lady with her newborn in the nest box.

Snow Leopard

Aptly nicknamed "ghost of the mountains," the snow leopard is perhaps the most beautiful and certainly the most mysterious of all the big cats.

Because they live in such difficult terrain in the wild — the rugged Himalayan plateau, the high country of Mongolia and Pakistan — studying them has been almost impossible. Finally, the placement of motion-sensitive cameras has provided scientists with a wealth of snow leopard images to study.

Black-necked swan and cygnet

The baby koala at the Cleveland Zoo. I think this says it all.

■ ■ ■ ■

KIDS WILL
BE KIDS

■ ■ ■ ■

Mountain lion cubs have spots when they're young.

No matter what the species, young-sters love to play. . . .

BABY PENINSULAR PRONGHORNS

These newborns still have to get their legs sorted out.

The peninsular pronghorn is another animal, like our California condor, that we are trying to bring back from a critically endangered status. Peninsular pronghorns are endangered. Once numbering in the thousands, there are now approximately two hundred and fifty in the wild. Since 2000, the Los Angeles Zoo has participated in the Peninsular Pronghorn Recovery Project. Our breeding herd is a collaboration between our zoo and several other zoos and Mexican wildlife organizations.

These baby pronghorns act just like the hungry kids they are.

Our Glenda in a fun mood.

Even the remote snow leopard is prone to occasional attacks of adorableness.

You learn it as a kid!

■ ■ ■ ■

AIRBORNE

■ ■ ■ ■

A pair of macaws in flight

From our feathered to our furry friends, the sight of an animal in mid-air is breathtaking.

CHEEKS THE COCKATOO

The L.A. Zoo has a great bird show that starts with a keeper coming out and asking, "Anyone have a dollar?"

Everyone takes a bill out and waves it in the air, and suddenly Cheeks, who is a free-flying cockatoo, comes out of nowhere and takes it from the hand of one of the visitors. It's a thrilling moment. He circles then redelivers the dollar to its owner to start the show. Here I am, having my money returned.

Red-tailed hawk

Of course, you don't really need feathers.

Speke's gazelle

Markhor kids

Sea lion

Baby giraffe

■ ■ ■ ■

CAMOUFLAGE

■ ■ ■ ■

Ornate horned toad

The marvelous array of colors and shapes and stripes and patterns that each animal wears is specifically designed by Mother Nature to help that animal blend in to his environment and virtually disappear.

It is a matter of survival . . . whether that be as the predator or as the prey.

The following are some extraordinary examples.

African Wild Dogs

African wild dogs probably take the prize as masters of camouflage.

Also known as cape hunting dogs or painted wolves, these animals are not domestic dogs gone wild, nor are they wolves, nor hyenas. They are a genus all their own. Unfortunately, their problem is familiar — their native habitat in Africa is growing smaller, and their population is following suit.

Observing these animals at the zoo is great if they are standing on grass, but, as you can see, if they are against the rocks, they become virtually invisible.

Where is he? With that broken pattern, he can disappear against almost any background. . . .

TIGERS

Why do tigers have stripes?

Wouldn't it seem that a huge orange predator with vivid black stripes would be easy enough to spot in time to frighten off potential prey? However, in the changing lights and shadows of the jungle, a tiger can almost disappear.

Despite their remarkable camouflage, the number of these magnificent animals in the wild continues to dwindle alarmingly. Save the tiger, indeed.

As you can see, even in a zoo environment, our beautiful Sumatran tiger manages to play his own game of hide-and-seek.

GREVY'S ZEBRAS

Meet Frances, Lewa, and Akina, three members of our group of Grevy's zebras.

In the wild, due to loss of habitat, competition with livestock, and poaching, these represent another diminishing population. Former vast herds have come down to less than two thousand animals in their natural African environment.

Grevy's are easily distinguished from other varieties of zebras by their close, narrow stripes and their pure white bellies, but telling individual animals apart can be a challenge unless you know that no two zebras are marked exactly alike. Like human fingerprints — or snowflakes. Look closely at these beauties and you'll see the subtle differences in their stripe patterns. When these guys get in motion, when they're running, they just disappear.

Growing up, I used to wonder who checked all the snowflakes to know that no two were alike. I'll take their word on zebras.

No, the stripes won't come off.

CHAMELEON

Camouflage is his business.

Bongo with his zoo friend — a yellow-backed duiker.

THE BONGO

The bongo is an antelope, and ours is the eastern mountain bongo . . . very rare and found only in the mountains of Kenya. The L.A. Zoo got its first bongo in 1975.

The day our bongo arrived, Dr. Thomas called me and asked, "Would you like to meet him?" I was astounded, because I knew the animal had to go into a period of quarantine before it could be put on exhibit. Of course I couldn't get there fast enough.

Warren took me backstage, and I

met our gorgeous new guy. As I petted him, Warren said, "Put your hand on his side." I did . . . and realized "he" was pregnant!

We had our bongo — two for the price of one!

In nature, these animals live in wooded mountains where the bright color may not seem the best camouflage. But when he's in motion, it's no problem.

More important — especially to him — it is his color that attracts the females.

Our jaguar in dappled sunshine.

■ ■ ■ ■

SLEEPYHEADS

■ ■ ■ ■

Sometimes zoo visitors are disappointed when they come to an exhibit to find the animals asleep, and they quickly move on to the next exhibit. If you want to be sure to see a wide-awake animal, you should get to the zoo early. So many animals are nocturnal hunters or gatherers. In the wild, they would sleep in the day as well.

Of course there are those of us zoo nuts who find animals fascinating even when they are snoozing. It can be a great opportunity to see them in detail without distraction. It may also be a good time to talk to the keepers and discover the deep bond that can exist between these caretakers and their charges.

Meet some of our sleepyheads.

HARRIET

What looks like a comfortable nap here would be far too dangerous in the wild for this youngster. Perhaps "comfortable" is not the right word, as it takes quite a bit of folding to get into this position, but here at the zoo, our Harriet has no need to worry about predators as she snoozes.

We even sleep adorably.

■ ■ ■ ■

NOSE
TO
NOSE

■ ■ ■ ■

This category really speaks for itself. . . .

Flamingos

Nubian ibex

Daddy takin and baby

Aldabra tortoises

Okapi

An apology is necessary after my visit to the Cleveland Zoo:
I'm sorry I got lipstick on your giraffe.

And to the Sacramento Zoo: Ditto!

Koko

Some of your close friends are hard to explain. This is Koko, the famous signing gorilla. In the background is her mentor, Dr. Penny Patterson.

Harbor seal at Georgia Aquarium

■ ■ ■ ■

My Misunderstood
Friends

■ ■ ■ ■

Three-banded armadillo

I remember the first time I ever saw a snake.

We were packing into the High Sierras on horseback. My dad was absolutely petrified of snakes. He could look at them and just go bananas. Well, the very first time we made a pack trip into the mountains I was four years old. I rode on his horse in front of him, my legs sticking straight out the sides. We were riding along these switchback trails over the passes, and as we came back down the other side of one particular mountain, here's this snake coming up the path. I was fascinated. My father, however, nearly jumped off — no, he stayed <u>on</u> — the horse, for good reason. I vividly remember that moment and my father's panic.

Later in my childhood we did a driving vacation through Idaho, and I remember stopping to look at the view of a beautiful meadow, and my father said, "There are thousands of snakes out there!" And I had visions of parting the grass and seeing it teeming. There were probably one or two out there, but to him there were "thousands."

I have close friends who want nothing to do with reptiles: Donna Ellerbusch, Patty Sullivan, and Loretta Barrett. However, when they saw how relaxed and beautiful these creatures are, they got terribly brave and actually held them and felt completely different about them!

This chapter isn't about only snakes and reptiles — there are many animal friends you might just learn to love if you give them a chance.

I was lucky. I didn't inherit my father's fear of snakes or of any type of animal — except maybe humans.

JACOB THE BOA CONSTRICTOR

Each spring, the zoo holds its fundraiser, the Beastly Ball, which is fantastic. When you arrive, you walk down a path where there are various keepers holding animals — most of whom you can't touch.

And then there's Jacob, the boa constrictor. Yes, I am close friends with a boa constrictor. Jacob loves to be held. He slips into your arms, where he relaxes completely.

This is an Aruba rattlesnake. Once again, the picture says it all.

Rowley's palm viper

BABY EMERALD TREE BOA

This is a baby emerald tree boa. The way he's sitting is a natural position for a snake — he can unwind and have purchase but still shoot forward to grab his prey.

These snakes seem languid in the day, but they're nocturnal. So at night the reptile house is alive with movement.

ARMADILLO

Uncurled, you get a whole armadillo. But when he curls up, he's perfectly spherical and all you get is a hard little armored ball.

Otis

MAGGIE
AND OTIS

When I first began working with the L.A. Zoo, Maggie and Otis were already established residents. A mated pair of hippopotamuses — hippopotami? — they were devoted to each other.

There are only two species of hippopotamus existing today — the pygmy hippo and the common, or Nile, hippo. The name "hippopotamus" comes from the ancient Greek word for water horse, but the hippo's closest relative is the whale.

Maggie and Otis weighed in somewhere in the eight-thousand-pound neighborhood. Pygmy hippos are much smaller — about three hundred and fifty pounds. Pygmy hippos birth their young on land, but their large common cousins have their babies underwater. Over the years I went through three birthings with Maggie and was always thrilled to see a new miniature version swimming after its mommy.

Maggie

ARABIAN ORYX

In profile, this animal looks to be one-horned and may have started the legend of the unicorn.

TORTOISE

The giant tortoise can live for more than one hundred years.

ROCK HYRAX

These little mammals are called rock hyrax. Believe it or not, they are the closest living relative of the elephant!

KOMODO DRAGONS

These spectacular creatures are the largest living lizards on earth — and are carnivorous. They will attack prey larger than themselves. There are about four to five thousand living in the wild, in Indonesia, where they are vulnerable, according to the International Union for Conservation of Nature. These guys are impressive to see, and with a name that includes the word "dragon," they are a great attraction in zoos. Actually, they have been somewhat rare in captivity, because they contract infectious and

parasitic diseases easily and suppos-
edly do not reproduce too readily.

Well, this is certainly not the case at
the L.A. Zoo, where, for the second
year in a row, we have a batch of more
than twenty eggs incubating. Last
year, twenty-two of twenty-three eggs
hatched viable live Komodos!

My question: Why are they so rare
when they have so many babies?

MIDORI THE MOUNTAIN TAPIR

The mountain tapir comes from Colombia and Ecuador, where they're sometimes kept as pets. They're an interesting combination — related to both the horse and the rhinoceros, due to the fact that they have an odd number of toes. They're very rare in captivity — there are only twelve of them at five zoos around the world. They're incredibly gentle — if you scratch them under their chins, they're completely sedated and the keepers can administer medicine or do tests on them if needed. Tapirs have a gestation period of thirteen months. Our Midori has had eight babies! Her firstborn was sent to another zoo in the U.S., and then to a Canadian zoo. Another of her offspring was sent to Colombia. In these ways, we can educate others about animals from around the world — and those who live in the local habitat as well. This helps an endangered species like the tapir, whose predators are jaguars . . . and humans.

■ ■ ■ ■

CONCLUSION

■ ■ ■ ■

Thank you for taking this little zoo tour with me. Hope you had a good time. I know I did.

On your next trip to your zoo, please think of us and look a little deeper. If you see something you don't like, report it, of course, but really enjoy the things you <u>do</u> like, and spread the word.

Love,

Betty

When I lost Allen, I was so deeply moved that the Koala Pavilion was dedicated to him with a plaque showing his great love of animals. There was another wonderful friend involved in making that possible: Grant Tinker.

■■■■

PRECIOUS RESOURCES

■■■■

TAD MOTOYAMA

Tad Motoyama and I first met when he came to work at the L.A. Zoo as official photographer in 1986. Over time he became aware of my passion for animals and began giving me prints of his incredible photos, which, of course, I have treasured.

At long last it dawned on me that these works of art must be shared. Hence this book.

A California native, Tad was born in 1942 in Manzanar, the Japanese internment facility, during World War II. His interest in photography began in high school.

Tad has a remarkable eye that enables him to capture images of animal behavior at crucial moments that we often miss.

A true artist.

ACKNOWLEDGMENTS

In all my years at the zoo, so many people contributed to this project that it would be impossible to name them. However, I must say a special thank-you to my longtime friend and literary agent, Loretta Barrett, to my new <u>good</u> friend and publisher at Putnam, Marysue Rucci, and to Jason Jacobs, head of publicity at the Los Angeles Zoo, for making it all possible. I thank all concerned from the bottom of my heart.

WEBSITES FOR FURTHER INFORMATION AND DISCOVERY

Listed below are the websites for the zoos and aquariums with whom we worked in the development of this book. For comprehensive listings of accredited zoos and aquariums, and to find your local zoo, please visit the website for the Association of Zoos and Aquariums at aza.org, or the World Association of Zoos and Aquariums at waza.org

Los Angeles Zoo and Botanical Gardens
www.lazoo.org

Sacramento Zoo
www.saczoo.org

Cleveland Metroparks Zoo
www.clemetzoo.com

Georgia Aquarium
www.georgiaaquarium.org

Albuquerque BioPark Zoo
www.cabq.gov/biopark/zoo

The Gorilla Foundation
www.koko.org

Columbus Zoo and Aquarium
www.colszoo.org

Zoo Atlanta
www.zooatlanta.org

San Diego Zoo
www.sandiegozoo.org

Brookfield Zoo
www.brookfieldzoo.org

Lincoln Park Zoo
www.lpzoo.org

Central Park Zoo
www.centralparkzoo.com

Philadelphia Zoo
www.philadelphiazoo.org

Denver Zoo
www.denverzoo.org

Woodland Park Zoo
www.zoo.org

Endangered Wolf Center
www.endangeredwolfcenter.org

The Living Desert
www.livingdesert.org

PHOTO CREDITS

Photos on pages 57, 81, 94, 97, 120, 132, 155, 190, 205, 208, 220, 278, 280, 288, and 325: Courtesy of Tad Motoyama

Photo on page 109: Courtesy of Betty White Private Collection

Photo on page 110: Courtesy of Neil Johnston

Photo on page 111: Courtesy of Jennifer Bell, Senior Zookeeper/ABQ BioPark

Photos on pages 112, 114, 256, and 270: Courtesy of Georgia Aquarium

Photos on pages 116 and 119: Courtesy of Dr. Thomas at the Los Angeles Zoo

Photos on pages 187 and 265: Courtesy of Cleveland Metroparks Zoo

Photo on page 266: Courtesy of Sacramento Zoo Staff Photo

Photo on page 268: Courtesy of Dr. Ron Cohn/Gorilla Foundation/koko.org

Photo on page 308: Courtesy of Miguel Gutierrez

All other photos: Courtesy of Tad Motoyama/Los Angeles Zoo

NEARLY THE END

THE END

ABOUT THE AUTHOR

Betty White has had an award-winning career in television and film, including iconic roles on *The Mary Tyler Moore Show* and *The Golden Girls.* She is also widely recognized for her lifelong work for animal welfare. She lives in Brentwood, California, with her golden retriever, Pontiac.

The employees of Thorndike Press hope you have enjoyed this Large Print book. All our Thorndike, Wheeler, and Kennebec Large Print titles are designed for easy reading, and all our books are made to last. Other Thorndike Press Large Print books are available at your library, through selected bookstores, or directly from us.

For information about titles, please call:

(800) 223-1244

or visit our Web site at:

http://gale.cengage.com/thorndike

To share your comments, please write:

Publisher
Thorndike Press
10 Water St., Suite 310
Waterville, ME 04901